CONVERSION FACTORS
SI UNITS AND MANY OTHERS

Over 2,100 conversion factors for biologists and mechanical engineers arranged in 21 quick-reference tables

Definitions - Dimensions - Abbreviations

C. J. Pennycuick

THE UNIVERSITY OF CHICAGO PRESS
Chicago & London

The University of Chicago Press, Chicago 60637
The University of Chicago Press, Ltd., London

Library of Congress Cataloging in Publication Data

Pennycuick, C. J. (Colin J.)
  [Handy matrices of unit conversion factors for biology and mechanics]
  Conversion factors : SI units and many others : over 2100 conversion factors for biologists and mechanical engineers arranged in 21 quick-reference tables, definitions, dimensions, abbreviations / C.J. Pennycuick. — 1988 ed.
     p.   cm.
  Reprint. Originally published: Handy matrices of unit conversion factors for biology and mechanics. New York : Wiley, c1974. With new introd.
  Includes index.
  1. Metric system—Conversion tables.   2. Biology—Tables.
3. Mechanics—Tables.   I. Title.
QC94.P46   1988   87-30145
530.8'12—dc19   CIP
ISBN 0-226-65507-5 (pbk.)

NOTE TO THE 1988 EDITION

This book is essentially a reprint of *Handy Matrices,* first published by Edward Arnold in 1974, and now out of print. I hope those readers of the earlier book, who wrote in with comments, will forgive me for reproducing the tables without revisions, in the interest of economy. I have revised the Introduction, but the rest of the book is the same as the original *Handy Matrices.* I am grateful to those colleagues and correspondents who have urged me and the publishers to get the book reprinted, and hope that this new University of Chicago Press edition will be useful to one and all.

C. J. Pennycuick
Miami, Florida, September 1987

CONTENTS

INTRODUCTION

A universal, internally consistent system of units has such obvious advantages that even Napoleon saw the point of it. However it is one thing to devise such a utopian system, as French scientists did two centuries ago, and quite another to persuade citizens, scientists or even government organisations to use it. For instance, it may come as a surprise to many Americans that the use of metric units has been mandatory in the United States since 1893. Acts of Congress notwithstanding, NASA charted its course to the moon and back in feet and pounds, while the American public continues to conduct its affairs in inches or miles, calories, British thermal units, gallons, horsepower and many other such numerical relics of the industrial revolution.

Since 1960, the "Système Internationale", or SI, has been the official standard for scientists everywhere, and most scientific journals urge their contributors to use SI units. However, old habits die hard. The calorie, which does not belong to any system of units, continues to be used by many physiologists and ecologists as though energy were some arbitrary form of currency, unrelated to force or distance. For them, James Joule might never have lived. As for time rates, they are measured per minute, per hour, per day or per year, hardly ever per second as the SI system requires. Scientists, like the public at large, prefer the familiar to the rational, and thereby too often obscure or overlook the underlying unity that links different areas of science. Engineers are fully as erratic as biologists in this respect, if not more so. Anyone using American pressure measuring equipment, for example, will soon feel the need to chop and change between pounds per square foot, pounds per square inch, inches of mercury, inches of water, and a few others. More insidiously, European engineers use a version of the metric system which looks superficially like the SI, but is in fact incompatible with it.

Physical and technical unit systems

It is not always appreciated that there are four different unit systems still in widespread use, each complete and internally consistent within itself. The "metric system" comes in physical and technical versions, which are different systems, incompatible with each other. The confusion is extreme, as the same unit names are used for quantities with different physical dimensions. The old British imperial system, which nowadays lingers mainly in the United States, likewise comes in physical and technical versions, in which the same unit names are used to represent different quantities.

To take a metric example, fluid density in the European engineering literature is commonly expressed in $kg\ s^2\ m^{-4}$. This mystifying unit is obviously based on the kilogramme, metre and second, but it is not an SI unit. The dimensions clearly do not correspond to density in the SI system. It belongs to the "technical" version of the kilogramme-metre-second system, as opposed to the "physical" version, which is the SI. In both versions, the kilogramme is defined by reference to a particular lump of platinum-iridium alloy, preserved in a kind of scientific shrine at Sèvres, France, but the definitions are based on different properties of this lump. In the physical (SI) system the kilogramme is defined as the mass of the lump, whereas in the technical system, it is the weight. Thus the technical system requires an assumed standard value for the acceleration due to gravity, whereas the physical system does not. The technical system also requires a different unit for mass, hence the "metric slug" or "technical metric mass unit". The density unit in the imperial technical system is the slug per cubic foot, which (improbable though this may sound to a biologist) is generally used for fluid density in the American, and older British, engineering literature.

In this book the names "kilogramme-force", "pound-force" etc, abbreviated "kgf", "lbf" etc, are used for the technical force units, to distinguish them from similarly named physical mass units. The value for the acceleration due to gravity now adopted by international agreement is 9.806650 m s^{-2}, or 32.17405 ft s^{-2}, and the conversion factors in this book are based on these values. Different standards for gravity have been in use at various times in the past, and in some British and American engineering literature rounded values of 32.17 or even 32.2 ft s^{-2} have been used. This is reflected in the conversion factors between corresponding physical and technical units, as between the pound and the slug, the pound-force and the poundal, etc.

Confusion between the technical and physical versions of a system is a common source of disastrous error, which can be avoided by paying attention to physical dimensions. In particular, many biologists seem chronically schizoid about weight and mass. They "weigh" an animal on a spring balance (which measures weight) or on a beam balance (which measures mass), and in either case most often express the result in mass units, but describe it as "weight". Usually the mass is meant, and the kilogramme is the appropriate SI unit. However, in those areas of biology where one actually needs to measure weights, as in locomotion studies, the SI unit of weight is the newton, not the kilogramme.

Objectives of this book
The first objective of this book is to provide a handy reference for those who need to find out what the SI unit actually is for grassland productivity, kinematic viscosity or whatever. Then, a set of units, each thought to have been used somewhere for scientific purposes, has been chosen for each of 21 different physical quantities. For each set, a table is given, showing the conversion factor for changing from each unit in the set to every one of the others. Finally, the physical dimensions for each quantity are given at the head of the table, in both the physical system (based on mass, length and time), and the technical system (based on force, length and time). The more intricate units, such as the metric technical density unit mentioned above, can be understood by inspection of these dimensions. The dimensions are also tabulated in the Table of Contents.

How to use the conversion tables
For each type of physical quantity, a main table is provided, in which the conversion factors are given in scaled form, consisting of a 5-digit number, followed by the power of 10 by which it is to be multiplied. For instance, to convert kilocalories per day into watts, the entry in the "Power" table is 4.8458; -2. This means "multiply by 4.8458×10^{-2}" in other words, by 0.048458. In most cases the conversion factor from any unit to any other can be found by looking up a single entry in the main table, but in some cases the conversion factors between different metric units (integral powers of 10) have been relegated to supplementary tables. Some conversions have to be done in two stages. For example, to convert a pressure expressed in tons-force per square inch into bars, first look up the conversion factor from tonf in^{-2} into pascals (the SI unit), and then from Pa into bars, thus.

tonf in^{-2} to Pa: 1.5444×10^{7}

Pa to bar: 10^{-5}

whence

tonf in^{-2} to bar: 1.5444×10^{2}.

In the case of force, and quantities derived from it, two supplementary tables are needed, one for physical and one for technical metric units. The unit to be used for the intermediate step is always the one in the first row and column of the supplementary table (top left corner).

Abbreviations of unit names

The names of the units are written out in full at the left-hand side of each table, and the columns are headed by the abbreviations for the same units, in the same order. Wherever possible the abbreviations given are the approved ones listed in British Standard 1991 (1967), or in B.S. 350 (1959), or follow the same conventions. In the case of units not mentioned in these publications, abbreviations known to be in common use are given. In a few cases, abbreviations have had to be invented, in which case the style of the approved abbreviations has been followed as closely as possible.

Sources and accuracy

Almost all the conversion factors in these tables have been calculated from the "key factors" given in British Standard 350 (1959). The tables are not completely internally consistent at the level of the least significant digit, because of rounding errors. To minimise these, each table was first calculated to an accuracy of 7 significant digits, and checked for internal consistency at that level. Finally, the factors were rounded to 5 significant digits. It is hoped that in this way the rounding errors have been kept within one unit of the fifth digit.

References

Further information on the principles of measurement and unit systems, and on the origins and definitions of particular units, can be found in the following works:

British Standard 350. *Conversion factors and tables. Part 1. Basis of tables. Conversion factors.* British Standards Institution (1959).

British Standard 3763. *The international system of units (SI).* British Standards Institution (1970).

British Standard PD 5686. *The use of SI units.* British Standards Institution (1972).

Eshbach, O.W. *Handbook of engineering fundamentals.* Wiley, New York (2nd edition, 1952).

Kaye, G.W.C. and Laby, T.H. *Tables of physical and chemical constants.* Longmans & Green, London (11th edition, 1956).

Mechtly, E.A. *The International System of units. Physical constants and conversion factors.* Second revision, NASA SP-7012 (1973).

In addition, a comprehensive and fascinating list of weights and measures used locally in different parts of the world, with the metric equivalents, is given in:

U.N. Publication ST/STAT/SER.M/21/Rev. 1. *World weights and measures. Handbook for statisticians.* United Nations Department of Economic and Social Affairs, New York (1966).

METRIC PREFIXES AND MULTIPLIERS

Prefix	Abbreviation	Multiplier
tera-	T	10^{12}
giga-	G	10^9
mega-	M	10^6
kilo-	k	10^3
hecto-	h	10^2
deca-	da	10
deci-	d	10^{-1}
centi-	c	10^{-2}
milli-	m	10^{-3}
micro-	μ	10^{-6}
nano-	n	10^{-9}
pico-	p	10^{-12}
femto-	f	10^{-15}
atto-	a	10^{-18}

LENGTH   DISTANCE

Dimensions   L

SI Unit   metre (m) (arbitrarily defined)

Notes   One *nautical mile* on the earth's surface subtends an angle of approximately one arc minute at the centre of the earth. The British Admiralty nautical mile is equal to 6080 ft, and is about 1.19 m longer than the international unit.

The *astronomical unit* is the mean distance of the earth from the sun. Until recently it was known only to 4-figure accuracy, but its accepted value has now been refined to 1.495979×10^{11} m as a result of interplanetary radar studies (reviewed by M. E. Ash, I. I. Shapiro and W. B. Smith (1967) *Astron. J.* **72**, 338–350).

The *parsec* is the distance at which 1 au subtends an angle of 1 arc second.

The *light year* given here is based on the tropical year 1900.

Into ↓ *To convert* →	m	in	thou (= mil)	ft	yd	fath	rod	ch
metre	1	2.5400; −2	2.5400; −5	3.0480; −1	9.1440; −1	1.8288; 0	5.0292; 0	2.0117; +1
inch	3.9370; +1	1	1.0000; −3	1.2000; +1	3.6000; +1	7.2000; +1	1.9800; +2	7.9200; +2
thousandth of an inch (= mil)	3.9370; +4	1.0000; +3	1	1.2000; +4	3.6000; +4	7.2000; +4	1.9800; +5	7.9200; +5
foot	3.2808; 0	8.3333; −2	8.3333; −5	1	3.0000; 0	6.0000; 0	1.6500; +1	6.6000; +1
yard	1.0936; 0	2.7778; −2	2.7778; −5	3.3333; −1	1	2.0000; 0	5.5000; 0	2.2000; +1
fathom	5.4681; −1	1.3889; −2	1.3889; −5	1.6667; −1	5.0000; −1	1	2.7500; 0	1.1000; +1
rod	1.9884; −1	5.0505; −3	5.0505; −6	6.0606; −2	1.8182; −1	3.6364; −1	1	4.0000; 0
chain	4.9710; −2	1.2626; −3	1.2626; −6	1.5152; −2	4.5455; −2	9.0909; −2	2.5000; −1	1
furlong	4.9710; −3	1.2626; −4	1.2626; −7	1.5152; −3	4.5455; −3	9.0909; −3	2.5000; −2	1.0000; −1
statute mile	6.2137; −4	1.5783; −5	1.5783; −8	1.8939; −4	5.6818; −4	1.1364; −3	3.1250; −3	1.2500; −2
international nautical mile	5.3996; −4	1.3715; −5	1.3715; −8	1.6458; −4	4.9374; −4	9.8747; −4	2.7156; −3	1.0862; −2
astronomical unit	6.6846; −12	1.6979; −13	1.6979; −16	2.0375; −12	6.1124; −12	1.2225; −11	3.3618; −11	1.3447; −10
parsec	3.2408; −17	8.2316; −19	8.2316; −22	9.8779; −18	2.9634; −17	5.9267; −17	1.6299; −16	6.5194; −16
light year	1.0570; −16	2.6848; −18	2.6848; −21	3.2218; −17	9.6654; −17	1.9331; −16	5.3160; −16	2.1264; −15

multiply by the factor in the appropriate cell of the table

metric units

Into ↓ To convert →	m	Å	µm	mm	cm	km	Nd mile
metre	0	−10	−6	−3	−2	+3	+4
ångstrom	+10	0	+4	+7	+8	+13	+14
micrometre (= micron)	+6	−4	0	+3	+4	+9	+10
millimetre	+3	−7	−3	0	+1	+6	+7
centimetre	+2	−8	−4	−1	0	+5	+6
kilometre	−3	−13	−9	−6	−5	0	+1
Nordic mile	−4	−14	−10	−7	−6	−1	0

multiply by 10^n, where n is the number in the appropriate cell of the table

fg	mile	NM	au	pc	lt y	← To convert Into ↓
2.0117; +2	1.6093; +3	1.8520; +3	1.4960; +11	3.0857; +16	9.4605; +15	metre
7.9200; +3	6.3360; +4	7.2913; +4	5.8897; +12	1.2148; +18	3.7246; +17	inch
7.9200; +6	6.3360; +7	7.2913; +7	5.8897; +15	1.2148; +21	3.7246; +20	thousandth of an inch (= mil)
6.6000; +2	5.2800; +3	6.0761; +3	4.9081; +11	1.0124; +17	3.1038; +16	foot
2.2000; +2	1.7600; +3	2.0254; +3	1.6360; +11	3.3745; +16	1.0346; +16	yard
1.1000; +2	8.8000; +2	1.0127; +3	8.1801; +10	1.6873; +16	5.1731; +15	fathom
4.0000; +1	3.2000; +2	3.6825; +2	2.9746; +10	6.1355; +15	1.8811; +15	rod
1.0000; +1	8.0000; +1	9.2062; +1	7.4365; +9	1.5339; +15	4.7028; +14	chain
1	8.0000; 0	9.2062; 0	7.4365; +8	1.5339; +14	4.7028; +13	furlong
1.2500; −1	1	1.1508; 0	9.2956; +7	1.9174; +13	5.8785; +12	statute mile
1.0862; −1	8.6898; −1	1	8.0776; +7	1.6661; +13	5.1083; +12	international nautical mile
1.3447; −9	1.0758; −8	1.2380; −8	1	2.0626; +5	6.3240; +4	astronomical unit
6.5194; −15	5.2155; −14	6.0019; −14	4.8481; −6	1	3.0659; −1	parsec
2.1264; −14	1.7011; −13	1.9576; −13	1.5813; −5	3.2616; 0	1	light year

AREA

Dimensions L^2

SI Unit square metre (m^2)

Notes The *circular inch* is the area of a circle whose diameter is one inch, and similarly with the *circular mil*.

The *square mile* is based on the statute mile.

The *morgen* is the unit used in South Africa, which differs slightly from several other units of the same name, formerly in use in various parts of Germany.

To convert → Into ↓	m^2	ci in	ci mil	in^2	ft^2	yd^2	rod^2	ch^2
square metre	1	5.0671; −4	5.0671; −10	6.4516; −4	9.2903; −2	8.3613; −1	2.5293; +1	4.0469; +2
circular inch	1.9735; +3	1	1.0000; −6	1.2732; 0	1.8335; +2	1.6501; +3	4.9916; +4	7.9866; +5
circular mil	1.9735; +9	1.0000; +6	1	1.2732; +6	1.8335; +8	1.6501; +9	4.9916; +10	7.9866; +11
square inch	1.5500; +3	7.8540; −1	7.8540; −7	1	1.4400; +2	1.2960; +3	3.9204; +4	6.2726; +5
square foot	1.0764; +1	5.4542; −3	5.4542; −9	6.9444; −3	1	9.0000; 0	2.7225; +2	4.3560; +3
square yard	1.1960; 0	6.0602; −4	6.0602; −10	7.7160; −4	1.1111; −1	1	3.0250; +1	4.8400; +2
square rod	3.9537; −2	2.0034; −5	2.0034; −11	2.5508; −5	3.6731; −3	3.3058; −2	1	1.6000; +1
square chain	2.4711; −3	1.2521; −6	1.2521; −12	1.5942; −6	2.2957; −4	2.0661; −3	6.2500; −2	1
rood	9.8842; −4	5.0084; −7	5.0084; −13	6.3769; −7	9.1827; −5	8.2645; −4	2.5000; −2	4.0000; −1
acre	2.4711; −4	1.2521; −7	1.2521; −13	1.5942; −7	2.2957; −5	2.0661; −4	6.2500; −3	1.0000; −1
morgen	1.1675; −4	5.9158; −8	5.9158; −14	7.5322; −8	1.0846; −5	9.7618; −5	2.9529; −3	4.7247; −2
square mile	3.8610; −7	1.9564; −10	1.9564; −16	2.4910; −10	3.5870; −8	3.2283; −7	9.7656; −6	1.5625; −4
square nautical mile	2.9155; −7	1.4773; −10	1.4773; −16	1.8810; −10	2.7086; −8	2.4378; −7	7.3742; −6	1.1799; −4

multiply by the factor in the appropriate cell of the table

metric units

To convert →	m^2	mm^2	cm^2	a	ha	km^2
Into ↓						
square metre	0	−6	−4	+2	+4	+6
square millimetre	+6	0	+2	+8	+10	+12
square centimetre	+4	−2	0	+6	+8	+10
are	−2	−8	−6	0	+2	+4
hectare	−4	−10	−8	−2	0	+2
square kilometre	−6	−12	−10	−4	−2	0

multiply by 10^n, where n is the number in the appropriate cell of the table

rood	acre	mo	$mile^2$	NM^2	← *To convert* / *Into* ↓
1.0117; +3	4.0469; +3	8.5053; +3	2.5900; +6	3.4299; +6	square metre
1.9966; +6	7.9866; +6	1.6904; +7	5.1114; +9	6.7690; +9	circular inch
1.9966; +12	7.9866; +12	1.6904; +13	5.1114; +15	6.7690; +15	circular mil
1.5682; +6	6.2726; +6	1.3276; +7	4.0145; +9	5.3164; +9	square inch
1.0890; +1	4.3560; +4	9.2061; +4	3.7878; +7	3.6910; +7	square foot
1.2100; +3	4.8400; +3	1.0244; +4	3.0976; +6	4.1021; +6	square yard
4.0000; +1	1.6000; +2	3.3865; +2	1.0240; +5	1.3561; +5	square rod
2.5000; 0	1.0000; +1	2.1165; +1	6.4000; +3	8.4755; +3	square chain
1	4.0000; 0	8.4661; 0	2.5600; +3	3.3902; +3	rood
2.5000; −1	1	2.1165; 0	6.4000; +2	8.4755; +2	acre
1.1812; −1	4.7247; −1	1	3.0238; +2	4.0044; +2	morgen
3.9063; −4	1.5625; −3	3.3071; −3	1	1.3243; 0	square mile
2.9497; −4	1.1799; −3	2.4972; −3	7.5512; −1	1	square nautical mile

AREA

VOLUME

Dimensions L^3

SI Unit   cubic metre (m^3)

Notes   The *litre* is defined as the volume occupied by 1 kg of pure water at its maximum density. It is just distinguishable from 10^{-3} m^3 at the 5-figure level of accuracy. The apparent discrepancy between the conversion factor from l to m^3 and its reciprocal is due to rounding errors. According to NASA SP-7012 the litre has been redefined as being exactly equal to 10^{-3} m^3, but this practice does not seem to be followed elsewhere.

To convert → Into ↓	m^3	l	in^3	ft^3	yd^3	$mile^3$	UK fl oz	UK pt
cubic metre	1	1.0000; −3	1.6387; −5	2.8317; −2	7.6455; −1	4.1682; +9	2.8413; −5	5.6826; −4
litre	9.9997; +2	1	1.6387; −2	2.8316; +1	7.6453; +2	4.1681; +12	2.8412; −2	5.6825; −1
cubic inch	6.1024; +4	6.1025; +1	1	1.7280; +3	4.6656; +4	2.5436; +14	1.7339; 0	3.4677; +1
cubic foot	3.5315; +1	3.5316; −2	5.7870; −4	1	2.7000; +1	1.4720; +11	1.0034; −3	2.0068; −2
cubic yard	1.3080; 0	1.3080; −3	2.1433; −5	3.7037; −2	1	5.4518; +9	3.7163; −5	7.4326; −4
cubic mile	2.3991; −10	2.3992; −13	3.9315; −15	6.7936; −12	1.8343; −10	1	6.8167; −15	1.3633; −13
UK fluid ounce	3.5195; +4	3.5196; +1	5.7674; −1	9.9661; +2	2.6909; +4	1.4670; +14	1	2.0000; +1
UK pint	1.7598; +3	1.7598; 0	2.8837; −2	4.9831; +1	1.3454; +3	7.3350; +12	5.0000; −2	1
UK gallon	2.1997; +2	2.1998; −1	3.6047; −3	6.2288; 0	1.6818; +2	9.1687; +11	6.2500; −3	1.2500; −1
US fluid ounce	3.3814; +4	3.3815; +1	5.5411; −1	9.5751; +2	2.5853; +4	1.4094; +14	9.6076; −1	1.9215; +1
US pint	2.1134; +3	2.1134; 0	3.4632; −2	5.9844; +1	1.6158; +3	8.8089; +12	6.0047; −2	1.2009; 0
US gallon	2.6417; +2	2.6418; −1	4.3290; −3	7.4805; 0	2.0197; +2	1.1011; +12	7.5059; −3	1.5012; −1

multiply by the factor in the appropriate cell of the table

UK gal		US fl oz		US pt		US gal		← *To convert* Into ↓
4.5461;	−3	2.9574;	−5	4.7318;	−4	3.7854;	−3	cubic metre
4.5460;	0	2.9573;	−2	4.7316;	−1	3.7853;	0	litre
2.7742;	+2	1.8047;	0	2.8875;	+1	2.3100;	+2	cubic inch
1.6054;	−1	1.0444;	−3	1.6710;	−2	1.3368;	−1	cubic foot
5.9461;	−3	3.8681;	−5	6.1889;	−4	4.9511;	−3	cubic yard
1.0907;	−12	7.0951;	−15	1.1352;	−13	9.0817;	−13	cubic mile
1.6000;	+2	1.0408;	0	1.6653;	+1	1.3323;	+2	UK fluid ounce
8.0000;	0	5.2042;	−2	8.3267;	−1	6.6614;	0	UK pint
1		6.5053;	−3	1.0408;	−1	8.3267;	−1	UK gallon
1.5372;	+2	1		1.6000;	+1	1.2800;	+2	US fluid ounce
9.6076;	0	6.2500;	−2	1		8.0000;	0	US pint
1.2009;	0	7.8125;	−3	1.2500;	−1	1		US gallon

TIME

Dimensions T

SI Unit second (s) (arbitrarily defined)

Notes The *lunar month* (or synodic month) is the time taken for the moon to return to the same phase.

The *sidereal month* is the time taken for the moon to return to the same right ascension.

The *year* is the tropical year 1900.

To convert → Into ↓	s	min	h	d	si d	wk	mon	si mon	y
second	1	6.0000; +1	3.6000; +3	8.6400; +4	8.6164; +4	6.0480; +5	2.5514; +6	2.3606; +6	3.1557; +7
minute	1.6667; −2	1	6.0000; +1	1.4400; +3	1.4361; +3	1.0080; +4	4.2524; +4	3.9343; +4	5.2595; +5
hour	2.7778; −4	1.6667; −2	1	2.4000; +1	2.3934; +1	1.6800; +2	7.0873; +2	6.5572; +2	8.7658; +3
mean solar day	1.1574; −5	6.9444; −4	4.1667; −2	1	9.9727; −1	7.0000; 0	2.9531; +1	2.7322; +1	3.6524; +2
sidereal day	1.1606; −5	6.9635; −4	4.1781; −2	1.0027; 0	1	7.0192; 0	2.9611; +1	2.7396; +1	3.6624; +2
week	1.6534; −6	9.9206; −5	5.9524; −3	1.4286; −1	1.4247; −1	1	4.2187; 0	3.9031; 0	5.2177; +1
lunar (= synodic) month	3.9194; −7	2.3516; −5	1.4110; −3	3.3863; −2	3.3771; −2	2.3704; −1	1	9.2520; −1	1.2368; +1
sidereal month	4.2362; −7	2.5417; −5	1.5250; −3	3.6601; −2	3.6501; −2	2.5621; −1	1.0808; 0	1	1.3368; +1
year	3.1689; −8	1.9013; −6	1.1408; −4	2.7379; −3	2.7304; −3	1.9165; −2	8.0852; −2	7.4804; −2	1

multiply by the factor in the appropriate cell of the table

SPEED

Dimensions	LT^{-1}
SI Unit	metre per second (m s^{-1})

Notes The *nautical mile* is the international nautical mile and the *day* is the mean solar day.

To convert → / Into ↓	m s⁻¹	cm s⁻¹	m min⁻¹	ft s⁻¹	ft min⁻¹	km h⁻¹	mile h⁻¹	kt	NM d⁻¹
metre per second	1	1.0000; −2	1.6667; −2	3.0480; −1	5.0800; −3	2.7778; −1	4.4704; −1	5.1444; −1	2.1435; −2
centimetre per second	1.0000; +2	1	1.6667; 0	3.0480; +1	5.0800; −1	2.7778; +1	4.4704; +1	5.1444; +1	2.1435; 0
metre per minute	6.0000; +1	6.0000; −1	1	1.8288; +1	3.0480; −1	1.6667; +1	2.6822; +1	3.0867; +1	1.2861; 0
foot per second	3.2808; 0	3.2808; −2	5.4681; −2	1	1.6667; −2	9.1134; −1	1.4667; 0	1.6878; 0	7.0325; −2
foot per minute	1.9685; +2	1.9685; 0	3.2808; 0	6.0000; +1	1	5.4681; +1	8.8000; +1	1.0127; +2	4.2195; 0
kilometre per hour	3.6000; 0	3.6000; −2	6.0000; −2	1.0973; 0	1.8288; −2	1	1.6093; 0	1.8520; 0	7.7167; −2
mile per hour	2.2369; 0	2.2369; −2	3.7282; −2	6.8182; −1	1.1364; −2	6.2137; −1	1	1.1508; 0	4.7949; −2
knot (= nautical mile per hour)	1.9438; 0	1.9438; −2	3.2397; −2	5.9248; −1	9.8747; −3	5.3996; −1	8.6898; −1	1	4.1667; −2
nautical mile per day	4.6652; +1	4.6652; −1	7.7754; −1	1.4220; +1	2.3699; −1	1.2959; +1	2.0855; +1	2.4000; +1	1

multiply by the factor in the appropriate cell of the table

MASS

Dimensions $\quad$ M or F L^{-1} T^2

SI Unit $\quad$ kilogramme (kg) (arbitrarily defined)

Notes $\quad$ The prefix T or P indicates that a unit belongs to a 'technical' (gravitational) or 'physical' system of units respectively (see Introduction).

	Into ↓ \ To convert →	kg	g	t	kgf s^2 m^{-1}	oz	lb	st
P	kilogramme	1	1.0000; −3	1.0000; +3	9.8067; 0	2.8350; −2	4.5359; −1	6.3503; 0
P	gramme	1.0000; +3	1	1.0000; +6	9.8067; +3	2.8350; +1	4.5359; +2	6.3503; +3
P	tonne	1.0000; −3	1.0000; −6	1	9.8067; −3	2.8350; −5	4.5359; −4	6.3503; −3
T	metric technical mass unit	1.0197; −1	1.0197; −4	1.0197; +2	1	2.8908; −3	4.6254; −2	6.4755; −1
P	ounce (avoirdupois)	3.5274; +1	3.5274; −2	3.5274; +4	3.4592; +2	1	1.6000; +1	2.2400; +2
P	pound	2.2046; 0	2.2046; −3	2.2046; +3	2.1620; +1	6.2500; −2	1	1.4000; +1
P	stone	1.5747; −1	1.5747; −4	1.5747; +2	1.5443; 0	4.4643; −3	7.1429; −2	1
P	quarter	7.8737; −2	7.8737; −5	7.8737; +1	7.7214; −1	2.2321; −3	3.5714; −2	5.0000; −1
P	hundredweight	1.9684; −2	1.9864; −5	1.9864; +1	1.9304; −1	5.5804; −4	8.9286; −3	1.2500; −1
P	UK ton (= long ton)	9.8421; −4	9.8421; −7	9.8421; −1	9.6518; −3	2.7902; −5	4.4643; −4	6.2500; −3
P	short ton	1.1023; −3	1.1023; −6	1.1023; 0	1.0810; −2	3.1250; −5	5.0000; −4	7.0000; −3
T	slug	6.8522; −2	6.8522; −5	6.8522; +1	6.7917; −1	1.9426; −3	3.1081; −2	4.3513; −1

multiply by the factor in the appropriate cell of the table

qr	cwt	UK ton	sh ton	slug	← To convert Into ↓
1.2701; +1	5.0802; +1	1.0160; +3	9.0718; +2	1.4594; +1	kilogramme
1.2701; +4	5.0802; +4	1.0160; +6	9.0718; +5	1.4594; +4	gramme
1.2701; −2	5.0802; −2	1.0160; 0	9.0718; −1	1.4594; −2	tonne
1.2951; 0	5.1804; 0	1.0361; +2	9.2507; +1	1.4882; 0	metric technical mass unit
4.4800; +2	1.7920; +3	3.5840; +4	3.2000; +4	5.1478; +2	ounce (avoirdupois)
2.8000; +1	1.1200; +2	2.2400; +3	2.0000; +3	3.2174; +1	pound
2.0000; 0	8.0000; 0	1.6000; +2	1.4286; +2	2.2981; 0	stone
1	4.0000; 0	8.0000; +1	7.1429; +1	1.1491; 0	quarter
2.5000; −1	1	2.0000; +1	1.7857; +1	2.8727; −1	hundredweight
1.2500; −2	5.0000; −2	1	8.9286; −1	1.4363; −2	UK ton (= long ton)
1.4000; −2	5.6000; −2	1.1200; 0	1	1.6087; −2	short ton
8.7027; −1	3.4811; 0	6.9621; +1	6.2162; +1	1	slug

DENSITY

To convert →	kg m⁻³	g m⁻³	g cm⁻³	mg cm⁻³	mg mm⁻³	μg mm⁻³
Into ↓						
kilogramme per cubic metre	0	−3	+3	0	+3	0
gramme per cubic metre	+3	0	+6	+3	+6	+3
gramme per cubic centimetre	−3	−6	0	−3	0	−3

Dimensions $M\,L^{-3}$ or $F\,L^{-4}\,T^{2}$

SI Unit kilogramme per cubic metre ($kg\,m^{-3}$)

multiply by 10^{n}, where n is the number in the appropriate cell of the table

To convert →	kg m⁻³	kg l⁻¹	kgf s² m⁻⁴	sl in⁻³	sl ft⁻³	sl yd⁻³	lb in⁻³	lb ft⁻³
Into ↓								
kilogramme per cubic metre	1	9.9997; +2	9.8067; 0	8.9057; +5	5.1538; +2	1.9088; +1	2.7680; +4	1.6018; +1
kilogramme per litre	1.0000; −3	1	9.8069; −3	8.9060; +2	5.1539; −1	1.9089; −2	2.7681; +1	1.6019; −2
metric technical mass unit per cubic metre	1.0197; −1	1.0197; +2	1	9.0813; +4	5.2554; +1	1.9464; 0	2.8226; +3	1.6334; 0
slug per cubic inch	1.1229; −6	1.1228; −3	1.1012; −5	1	5.7870; −4	2.1433; −5	3.1081; −2	1.7987; −5
slug per cubic foot	1.9403; −3	1.9403; 0	1.9028; −2	1.7280; +3	1	3.7037; −2	5.3708; +1	3.1081; −2
slug per cubic yard	5.2389; −2	5.2387; +1	5.1376; −1	4.6656; +4	2.7000; +1	1	1.4501; +3	8.3919; −1
pound per cubic inch	3.6127; −5	3.6126; −2	2.5429; −4	3.2174; +1	1.8619; −2	6.8960; −4	1	5.7870; −4
pound per cubic foot	6.2428; −2	6.2426; +1	6.1221; −1	5.5597; +4	3.2174; +1	1.1916; 0	1.7280; +3	1
pound per cubic yard	1.6856; 0	1.6855; +3	1.6530; +1	1.5011; +6	8.6870; +2	3.2174; +1	4.6656; +4	2.7000; +1
pound per UK gallon	1.0022; −2	1.0022; +1	9.8286; −2	8.9257; +3	5.1653; 0	1.9131; −1	2.7742; +2	1.6054; −1
pound per US gallon	8.3454; −3	8.3452; 0	8.1840; −2	7.4322; +3	4.3010; 0	1.5930; −1	2.3100; +2	1.3368; −1
ounce per cubic inch	5.7804; −4	5.7802; −1	5.6686; −3	5.1478; +2	2.9791; −1	1.1034; −2	1.6000; +1	9.2593; −3
ounce per cubic foot	9.9885; −1	9.9882; +2	9.7953; 0	8.8955; +5	5.1478; +2	1.9066; +1	2.7648; +4	1.6000; +1
UK ton per cubic yard	7.5248; −4	7.5246; −1	7.3793; −3	6.7014; +2	3.8781; −1	1.4363; −2	2.0829; +1	1.2054; −2

multiply by the factor in the appropriate cell of the table

To convert →	$kg\ m^{-3}$	$g\ m^{-3}$	$g\ cm^{-3}$	$mg\ cm^{-3}$	$mg\ mm^{-3}$	$\mu g\ mm^{-3}$
Into ↓						
milligramme per cubic centimetre	0	−3	+3	0	+3	0
milligramme per cubic millimetre	−3	−6	0	−3	0	−3
microgramme per cubic millimetre	0	−3	+3	0	+3	0

multiply by 10^n, where n is the number in the appropriate cell of the table

$lb\ yd^{-3}$	$lb\ (UK\ gal)^{-1}$	$lb\ (US\ gal)^{-1}$	$oz\ in^{-3}$	$oz\ ft^{-3}$	$UK\ ton\ yd^{-3}$	← To convert / Into ↓
5.9328; −1	9.9776; +1	1.1983; +2	1.7300; +3	1.0012; 0	1.3289; +3	kilogramme per cubic metre
5.9329; −4	9.9779; −2	1.1983; −1	1.7300; 0	1.0012; −3	1.3290; 0	kilogramme per litre
6.0497; −2	1.0174; +1	1.2219; +1	1.7641; +2	1.0209; −1	1.3551; +2	metric technical mass unit per cubic metre
6.6817, −7	1.1204; −4	1.3455; −4	1.9426; −3	1.1242; −6	1.4922, −3	slug per cubic inch
1.1511; −3	1.9360; −1	2.3250; −1	3.3567; 0	1.9426; −3	2.5786; 0	slug per cubic foot
3.1081; −2	5.2271; 0	6.2275; 0	9.0632; +1	5.2449; −2	6.9621; +1	slug per cubic yard
7.1490, 5	0.0017; 3	4.3290; −3	6.2500; −2	3.6169; −5	4.8011; −2	pound per cubic inch
3.7037; −2	6.2288; 0	7.4805; 0	1.0800; +2	6.2500; −2	8.2903, +1	pound per cubic foot
1	1.6818; +2	2.0197; +2	2.9160; +3	1.6875; 0	2.2400; +3	pound per cubic yard
5.9461; −3	1	1.2009; 0	1.7339; +1	1.0034; −2	1.3319; +1	pound per UK gallon
4.9511; −3	8.3267; −1	1	1.4438; +1	8.3550; −3	1.1091; +1	pound per US gallon
3.4294; −4	5.7674; −2	6.9264; −2	1	5.7870; −4	7.6818; −1	ounce per cubic inch
5.9259; −1	9.9661; +1	1.1969; +2	1.7280; +3	1	1.3274; +3	ounce per cubic foot
4.4643; −4	7.5080; −2	9.0167; −2	1.3018; 0	7.5335; −4	1	UK ton per cubic yard

BIOMASS DENSITY

Notes The biomass density of small aquatic organisms may sometimes be represented in the same way as ordinary density, as mass per unit volume. Even in this case, however, the standing crop is more often related to the surface through which the community receives its energy supply, than to the volume which it occupies. Living organisms are therefore usually considered to be distributed upon a 2-dimensional surface, hence the dimensions adopted here for biomass density.

Dimensions $\quad ML^{-2}$ or $FL^{-3} T^2$

SI unit $\quad$ kilogramme per square metre (kg m^{-2})

To convert →	kg m^{-2}	ton acre^{-1}	ton mile^{-2}	lb in^{-2}	lb ft^{-2}	lb yd^{-2}	lb acre^{-1}	lb mile^{-2}
Into ↓								
kilogramme per square metre	1	2.5107; −1	3.9230; −4	7.0307; +2	4.8824; 0	5.4249; −1	1.1209; −4	1.7513; −7
UK ton per acre	3.9829; 0	1	1.5625; −3	2.8003; +3	1.9446; +1	2.1607; 0	4.4643; −4	6.9754; −7
UK ton per square mile	2.5491; +3	6.4000; +2	1	1.7922; +6	1.2446; +4	1.3829; +3	2.8571; −1	4.4643; −4
pound per square inch	1.4223; −3	3.5711; −4	5.5798; −7	1	6.9444; −3	7.7160; −4	1.5942; −7	2.4910; −10
pound per square foot	2.0482; −1	5.1423; −2	8.0349; −5	1.4400; +2	1	1.1111; −1	2.2957; −5	3.5870; −8
pound per square yard	1.8433; 0	4.6281; −1	7.2314; −4	1.2960; +3	9.0000; 0	1	2.0661; −4	3.2283; −7
pound per acre	8.9218; +3	2.2400; +3	3.5000; 0	6.2726; +6	4.3560; +4	4.8400; +3	1	1.5625; −3
pound per square mile	5.7099; +6	1.4336; +6	2.2400; +3	4.0145; +9	2.7878; +7	3.0976; +6	6.4000; +2	1
ounce per square inch	2.2757; −2	5.7137; −3	8.9277; −6	1.6000; +1	1.1111; −1	1.2346; −2	2.5508; −6	3.9856; −9
ounce per square foot	3.2771; 0	8.2277; −1	1.2856; −3	2.3040; +3	1.6000; +1	1.7778; 0	3.6731; −4	5.7392; −7
ounce per square yard	2.9494; +1	7.4050; 0	1.1570; −2	2.0736; +4	1.4400; +2	1.6000; +1	3.3058; −3	5.1653; −6
gramme per square foot	9.2903; +1	2.3325; +1	3.6446; −2	6.5317; +4	4.5359; +2	5.0399; +1	1.0413; −2	1.6270; −5

multiply by the factor in the appropriate cell of the table

To convert →	kg m⁻²	kg km⁻²	kg ha⁻¹	t km⁻²	t ha⁻¹	g m⁻²	g cm⁻²	g mm⁻²	mg cm⁻²	mg mm⁻²
Into ↓										
kilogramme per square metre	0	−6	−4	−3	−1	−3	+1	+3	−2	0
kilogramme per square kilometre	+6	0	+2	+3	+5	+3	+7	+9	+4	+6
kilogramme per hectare	+4	−2	0	+1	+3	+1	+5	+7	+2	+4
tonne per square kilometre	+3	−3	−1	0	+2	0	+4	+6	+1	+3
tonne per hectare	+1	−5	−3	−2	0	−2	+2	+4	−1	+1
gramme per square metre	+3	−3	−1	0	+2	0	+4	+6	+1	+3
gramme per square centimetre	−1	−7	−5	−4	−2	−4	0	+2	−3	−1
gramme per square millimetre	−3	−9	−7	−6	−4	−6	−2	0	−5	−3
milligramme per square centimetre	+2	−4	−2	−1	+1	−1	+3	+5	0	+2
milligramme per square millimetre	0	−6	−4	−3	−1	−3	+1	+3	−2	0

multiply by 10^n, where n is the number of the appropriate cell of the table

oz in⁻¹	oz ft⁻²	oz yd⁻²	g ft⁻²	← To convert
				Into ↓
4.3942; +1	3.0515; −1	3.3906; −2	1.0764; −2	kilogramme per square metre
1.7502; +2	1.2154; 0	1.3504; −1	4.2872; −2	UK ton per acre
1.1201; +5	7.7796; +2	8.6429; +1	2.7438; +1	UK ton per square mile
6.2500; −2	4.3403; −4	4.8225; −5	1.5310; −5	pound per square inch
9.0000; 0	6.2500; −2	6.9444; −3	2.2046; −3	pound per square foot
8.1000; +1	5.6250; −1	6.2500; −2	1.9842; −2	pound per square yard
3.9204; +5	2.7225; +3	3.0250; +2	9.6033; +1	pound per acre
2.5091; +8	1.7424; +6	1.9360; +5	6.1461; +4	pound per square mile
1	6.9444; −3	7.7160; −4	2.4496; −4	ounce per square inch
1.4400; +2	1	1.1111; −1	3.5274; −2	ounce per square foot
1.2960; +3	9.0000; 0	1	3.1747; −1	ounce per square yard
4.0823; +3	2.8350; +1	3.1499; 0	1	gramme per square foot

BIOMASS DENSITY

MOMENT OF INERTIA

Dimensions $M L^2$ or $F L T^2$

SI Unit   kilogramme metre-squared (kg m^2)

Into ↓ \ To convert →	kg m^2	g cm^2	lb ft^2	lb in^2	sl ft^2	kgf s^2 m
kilogramme metre-squared	1	1.0000; −7	4.2140; −2	2.9264; −4	1.3558; 0	9.8067; 0
gramme centimetre-squared	1.0000; +7	1	4.2140; +5	2.9264; +3	1.3558; +7	9.8067; +7
pound foot-squared	2.3730; +1	2.3730; −6	1	6.9444; −3	3.2174; +1	2.3272; +2
pound inch-squared	3.4172; +3	3.4172; −4	1.4400; +2	1	4.6331; +3	3.3511; +4
slug foot-squared	7.3756; −1	7.3756; −8	3.1081; −2	2.1584; −4	1	7.2330; 0
kilogramme-force second-squared metre	1.0197; −1	1.0197; −8	4.2971; −3	2.9841; −5	1.3825; −1	1

multiply by the factor in the appropriate cell of the table

FORCE WEIGHT

Dimensions	$M\,L\,T^{-2}$ or F
SI Unit	newton (N)
	$1\,N = 1\,kg\,m\,s^{-2}$

that is, 1 N is that force which will impart to a mass of 1 kg an acceleration of $1\,m\,s^{-2}$.

		metric technical units			
To convert →	kgf (= kp)	tf	gf	mgf
Into ↓				
kilogramme-force (= kilopond)	0	+3	−3	−6
tonne-force	−3	0	−6	−9
gramme-force	+3	+6	0	−3
milligramme-force	+6	+9	+3	0

multiply by 10^n, where n is the number in the appropriate cell of the table

	To convert →	N	dyn	kgf (= kp)	tonf	lbf	ozf	pdl
	Into ↓							
P	newton	1	1.0000; −5	9.8067; 0	9.9640; +3	4.4482; 0	2.7801; −1	1.3825; −1
P	dyne	1.0000; +5	1	9.8067; +5	9.9640; +8	4.4482; +5	2.7801; +4	1.3825; +4
T	kilogramme-force (= kilopond)	1.0197; −1	1.0197; −6	1	1.0160; +3	4.5359; −1	2.8350; −2	1.4098; −2
T	ton-force	1.0036; −4	1.0036; −9	9.8421; −4	1	4.4643; −4	2.7902; −5	1.3875; −5
T	pound-force	2.2481; −1	2.2481; −6	2.2046; 0	2.2400; +3	1	6.2500; −2	3.1081; −2
T	ounce-force	3.5969; 0	3.5969; −5	3.5274; +1	3.5840; +4	1.6000; +1	1	4.9730; −1
P	poundal	7.2330; 0	7.2330; −5	7.0932; +1	7.2070; +4	3.2174; +1	2.0109; 0	1

multiply by the factor in the appropriate cell of the table

Notes   The prefix T or P indicates that a unit belongs to a 'technical' (gravitational) or 'physical' system of units respectively (see Introduction).

In the convention adopted here, the names of technical force units have the suffix 'force' (abbreviated '-f') to distinguish them from the corresponding mass units, whose names are unqualified. In much engineering literature, however, the names 'pound', 'kilogramme' etc. (abbreviated 'lb', 'kg' etc) are used without qualification to designate the *force* units. This is usually apparent from the context, but caution is needed.

It should be especially noted that in the SI system it is *not* permissible to express weights in kilogrammes. The appropriate unit is the newton.

MEMBRANE TENSION

Dimensions $M\,T^{-2}$ or $F\,L^{-1}$

SI Unit newton per metre ($N\,m^{-1}$)
$(1\ N\,m^{-1} = 1\ kg\,s^{-2})$

metric physical units

To convert → Into ↓	$N\,m^{-1}$	$N\,cm^{-1}$	$N\,mm^{-1}$	$dyn\,cm^{-1}$	$dyn\,mm^{-1}$	$dyn\,\mu m^{-1}$
newton per metre	0	+2	+3	−3	−2	+1
newton per centimetre	−2	0	+1	−5	−4	−1
newton per millimetre	−3	−1	0	−6	−5	−2
dyne per centimetre	+3	+5	+6	0	+1	+4
dyne per millimetre	+2	+4	+5	−1	0	+3
dyne per micrometre	−1	+1	+2	−4	−3	0

multiply by 10^{n}, where n is the number in the appropriate cell of the table

To convert → Into ↓	$N\,m^{-1}$	$dyn\,cm^{-1}$	$kgf\,m^{-1}$	$lbf\,in^{-1}$	$lbf\,ft^{-1}$	$pdl\,in^{-1}$	$pdl\,ft^{-1}$	$ozf\,in^{-1}$
newton per metre	1	1.0000; −3	9.8067; 0	1.7513; +2	1.4594; +1	5.4431; 0	4.5359; −1	1.0945; +1
dyne per centimetre	1.0000; +3	1	9.8067; +3	1.7513; +5	1.4594; +4	5.4431; +3	4.5359; +2	1.0945; +4
kilogramme-force per metre	1.0197; −1	1.0197; −4	1	1.7858; +1	1.4882; 0	5.5504; −1	4.6254; −2	1.1161; 0
pound-force per inch	5.7101; −3	5.7101; −6	5.5997; −2	1	8.3333; −2	3.1081; −2	2.5901; −3	6.2500; −2
pound-force per foot	6.8522; −2	6.8522; −5	6.7197; −1	1.2000; +1	1	3.7297; −1	3.1081; −2	7.5000; −1
poundal per inch	1.8372; −1	1.8372; −4	1.8017; 0	3.2174; +1	2.6812; 0	1	8.3333; −2	2.0109; 0
poundal per foot	2.2046; 0	2.2046; −3	2.1620; +1	3.8609; +2	3.2174; +1	1.2000; +1	1	2.4131; +1
ounce-force per inch	9.1362; −2	9.1362; −5	8.9596; −1	1.6000; +1	1.3333; 0	4.9730; −1	4.1441; −2	1
ounce-force per foot	1.0963; 0	1.0963; −3	1.0752; +1	1.9200; +2	1.6000; +1	5.9675; 0	4.9730; −1	1.2000; +1
ton-force per inch	2.5492; −6	2.5492; −9	2.4999; −5	4.4643; −4	3.7202; −5	1.3875; −5	1.1563; −6	2.7902; −5
ton-force per foot	3.0590; −5	3.0590; −8	2.9999; −4	5.3571; −3	4.4643; −4	1.6651; −4	1.3875; −5	3.3482; −4
ton-force per yard	9.1770; −5	9.1770; −8	8.9996; −4	1.6071; −2	1.3393; −3	4.9952; −4	4.1626; −5	1.0045; −3

multiply by the factor in the appropriate cell of the table

| metric technical units | | | | | | | | |
To convert → kgf m^{-1}	kgf cm^{-1}	kgf mm^{-1}	tf m^{-1}	tf cm^{-1}	gf cm^{-1}	gf mm^{-1}	mgf mm^{-1}	$\text{mgf }\mu\text{m}^{-1}$
Into ↓								
kilogramme-force per metre	0	+2	+3	+3	+5	−1	0	−3	0
kilogramme-force per centimetre	−2	0	+1	+1	+3	−3	−2	−5	−2
kilogramme-force per millimetre	−3	−1	0	0	+2	−4	−3	−6	−3
tonne-force per metre	−3	−1	0	0	+2	−4	−3	−6	−3
tonne-force per centimetre	−5	−3	−2	−2	0	−6	−5	−8	−5
gramme-force per centimetre	+1	+3	+4	+4	+6	0	+1	−2	+1
gramme-force per millimetre	0	+2	+3	+3	+5	−1	0	−3	0
milligramme-force per millimetre	+3	+5	+6	+6	+8	+2	+3	0	+3
milligramme-force per micrometre	0	+2	+3	+3	+5	−1	0	−3	0

multiply by 10^n, where n is the number in the appropriate cell of the table

ozf ft^{-1}	tonf in^{-1}	tonf ft^{-1}	tonf yd^{-1}	← To convert
				Into ↓
9.1212; −1	3.9228; +5	3.2690; +4	1.0897; +4	newton per metre
9.1212; +2	3.9228; +8	3.2690; +7	1.0097; +7	dyne per centimetre
9.3010; −1	4.0002; +4	3.3335; +3	1.1112; +3	kilogramme-force per metre
5.2083; −3	2.2400; +3	1.8667; +2	6.2222; +1	pound-force per inch
6.2500; −2	2.6880; +4	2.2400; +3	7.4667; +2	pound-force per foot
1.6757; −1	7.2070; +4	6.0058; +3	2.0019; +3	poundal per inch
2.0109; 0	8.6484; +5	7.2070; +4	2.4023; +4	poundal per foot
8.3333; −2	3.5840; +4	2.9867; +3	9.9556; +2	ounce-force per inch
1	4.3008; +5	3.5840; +4	1.1947; +4	ounce-force per foot
2.3251; −6	1	8.3333; −2	2.7778; −2	ton-force per inch
2.7902; −5	1.2000; +1	1	3.3333; −1	ton-force per foot
8.3705; −5	3.6000; +1	3.0000; 0	1	ton-force per yard

PRESSURE   STRESS

Dimensions $M L^{-1} T^{-2}$ or $F L^{-2}$

SI Unit   pascal (Pa)
$1 Pa = 1 N m^{-2}$

metric technical units

To convert → Into ↓	$kgf\,m^{-2}$	$kgf\,cm^{-2}$	$kgf\,mm^{-2}$	$tf\,m^{-2}$	$tf\,cm^{-2}$	$gf\,cm^{-2}$	$gf\,mm^{-2}$
kilogramme-force per square metre	0	+4	+6	+3	+7	+1	+3
kilogramme-force per square centimetre	−4	0	+2	−1	+3	−3	−1
kilogramme-force per square millimetre	−6	−2	0	−3	+1	−5	−3
tonne-force per square metre	−3	+1	+3	0	+4	−2	0
tonne-force per square centimetre	−7	−3	−1	−4	0	−6	−4
gramme-force per square centimetre	−1	+3	+5	+2	+6	0	+2
gramme-force per square millimetre	−3	+1	+3	0	+4	−2	0

multiply by 10^n, where n is the number in the appropriate cell of the table

To convert → Into ↓	$Pa\,(= N\,m^{-2})$	$kgf\,m^{-2}$ $(=mm\,H_2O)$	$lbf\,ft^{-2}$	$lbf\,in^{-2}$	$pdl\,ft^{-2}$	$pdl\,in^{-2}$	$tonf\,yd^{-2}$	$tonf\,ft^{-2}$	$tonf\,in^{-2}$	$in\,H_2O$
pascal (= newton per square metre)	1	9.8067; 0	4.7880; +1	6.8948; +3	1.4882; 0	2.1430; +2	1.1917; +4	1.0725; +5	1.5444; +7	2.4909; +2
kilogramme-force per square metre (= millimetre of water)	1.0197; −1	1	4.8824; 0	7.0307; +2	1.5175; −1	2.1852; +1	1.2152; +3	1.0937; +4	1.5749; +6	2.5400; +1
pound-force per square foot	2.0885; −2	2.0482; −1	1	1.4400; +2	3.1081; −2	4.4757; 0	2.4889; +2	2.2400; +3	3.2256; +5	5.2023; 0
pound-force per square inch	1.4504; −4	1.4223; −3	6.9444; −3	1	2.1584; −4	3.1081; −2	1.7284; 0	1.5556; +1	2.2400; +3	3.6127; −2
poundal per square foot	6.7197; −1	6.5898; 0	3.2174; +1	4.6331; +3	1	1.4400; +2	8.0078; +3	7.2070; +4	1.0378; +7	1.6738; +2
poundal per square inch	4.6665; −3	4.5762; −2	2.2343; −1	3.2174; +1	6.9444; −3	1	5.5609; +1	5.0049; +2	7.2070; +4	1.1624; 0
ton-force per square yard	8.3915; −5	8.2292; −4	4.0179; −3	5.7857; −1	1.2488; −4	1.7983; −2	1	9.0000; 0	1.2960; +3	2.0902; −2
ton-force per square foot	9.3239; −6	9.1436; −5	4.4643; −4	6.4286; −2	1.3875; −5	1.9981; −3	1.1111; −1	1	1.4400; +2	2.3225; −3
ton-force per square inch	6.4749; −8	6.3497; −7	3.1002; −6	4.4643; −4	9.6357; −8	1.3875; −5	7.7160; −4	6.9444; −3	1	1.6128; −5
inch of water	4.0146; −3	3.9370; −2	1.9222; −1	2.7680; +1	5.9744; −3	8.6032; −1	4.7842; +1	4.3058; +2	6.2003; +4	1
foot of water	3.3455; −4	3.2808; −3	1.6018; −2	2.3067; 0	4.9787; −4	7.1693; −2	3.9868; 0	3.5881; +1	5.1669; +3	8.3333; −2
millimetre of mercury	7.5006; −3	7.3556; −2	3.5913; −1	5.1715; +1	1.1162; −2	1.6073; 0	8.9384; +1	8.0445; +2	1.1584; +5	1.8683; 0
inch of mercury	2.9530; −4	2.8959; −3	1.4139; −2	2.0360; 0	4.3945; −4	6.3281; −2	3.5190; 0	3.1671; +1	4.5607; +3	7.3556; −2
standard atmosphere	9.8692; −6	9.6784; −5	4.7254; −4	6.8046; −2	1.4687; −5	2.1149; −3	1.1761; −1	1.0585; 0	1.5242; +2	2.4583; −3

multiply by the factor in the appropriate cell of the table

metric physical units

To convert →	Pa	bar	mbar	N cm⁻²	N mm⁻²	dyn cm⁻²	dyn mm⁻²
Into ↓							
pascal	0	+5	+2	+4	+6	−1	+1
bar	−5	0	−3	−1	+1	−6	−4
millibar	−2	+3	0	+2	+4	−3	−1
newton per square centimetre	−4	+1	−2	0	+2	−5	−3
newton per square millimetre	−6	−1	−4	−2	0	−7	−5
dyne per square centimetre	+1	+6	+3	+5	+7	0	+2
dyne per square millimetre	−1	+4	+1	+3	+5	−2	0

multiply by 10ⁿ, where n is the number in the appropriate cell of the table

ft H₂O	mm Hg	in Hg	atm	← To convert / Into ↓
2.9891; +3	1.3332; +2	3.3864; +3	1.0133; +5	pascal (= newton per square metre)
3.0480; +2	1.3595; +1	3.4532; +2	1.0332; +4	kilogramme-force per square metre (= millimetre of water)
6.2428; +1	2.7845; 0	7.0726; +1	2.1162; +3	pound-force per square foot
4.3353; −1	1.9337; −2	4.9115; −1	1.4696; +1	pound-force per square inch
2.0086; +3	8.9589; +1	1.1730; 0	9.9997; +4	poundal per square foot
1.3948; +1	6.2214; −1	1.5802; +1	4.7283; +2	poundal per square inch
2.5083; −1	1.1188; −2	2.8417; −1	8.5027; 0	ton-force per square yard
2.7870; −2	1.2431; −3	3.1574; −2	9.4474; −1	ton-force per square foot
1.9354; −4	8.6325; −6	2.1927; −4	6.5607; −3	ton-force per square inch
1.2000; +1	5.3524; −1	1.3595; +1	4.0678; +2	inch of water
1	4.4603; −2	1.1329; 0	3.3899; +1	foot of water
2.2420; +1	1	2.5400; +1	7.6000; +2	millimetre of mercury
8.8267; −1	3.9370; −2	1	2.9921; +1	inch of mercury
2.9500; −2	1.3158; −3	3.3421; −2	1	standard atmosphere

Notes Pressures expressed in terms of *water gauge* assume the density of water to be 1 kg l⁻¹, which is the case at 4°C.

Pressures in terms of *mercury gauge* assume the density of mercury to be 13.5955 kg l⁻¹, which is the case at 0°C. To convert the reading of a mercury manometer at T°C to the corresponding reading at 0°C, multiply by A, where

$$A = 1.00000 - (1.81453 \times 10^{-4})T + (2.19 \times 10^{-8})T^2$$

This equation has been calculated from the table given by Kaye and Laby (1956), relating the density of mercury to temperature (see p. 6 for full reference). For temperatures between −20°C and +100°C it predicts the values in the table to an accuracy better than ½ part in 10⁵. The correction takes account of the expansion of the mercury itself, but not that of the manometer vessel or scale.

WORK   ENERGY   HEAT   TORQUE

Dimensions $M\,L^2\,T^{-2}$ or FL
SI Unit   joule (J)
              $1\,J = 1\,N\,m$

Note   The *calorie* is the international table calorie.

To convert → Into ↓	J	erg	kgf m	gf cm	ft lbf	ft pdl	kcal	cal
joule	1	1.0000; −7	9.8067; 0	9.8067; −5	1.3558; 0	4.2140; −2	4.1868; +3	4.1868; 0
erg	1.0000; +7	1	9.8067; +7	9.8067; +2	1.3558; +7	4.2140; +5	4.1868; +10	4.1868; +7
kilogramme-force metre	1.0197; −1	1.0197; −8	1	1.0000; −5	1.3825; −1	4.2971; −3	4.2693; +2	4.2693; −1
gramme-force centimetre	1.0197; +4	1.0197; −3	1.0000; +5	1	1.3825; +4	4.2971; +2	4.2693; +7	4.2693; +4
foot pound-force	7.3756; −1	7.3756; −8	7.2330; 0	7.2330; −5	1	3.1081; −2	3.0880; +3	3.0880; 0
foot poundal	2.3730; +1	2.3730; −6	2.3272; +2	2.3272; −3	3.2174; +1	1	9.9354; +4	9.9354; +1
kilocalorie	2.3885; −4	2.3885; −11	2.3423; −3	2.3423; −8	3.2383; −4	1.0065; −5	1	1.0000; −3
calorie	2.3885; −1	2.3885; −8	2.3423; 0	2.3423; −5	3.2383; −1	1.0065; −2	1.0000; +3	1
kilowatt hour	2.7778; −7	2.7778; −14	2.7241; −6	2.7241; −11	3.7662; −7	1.1706; −8	1.1630; −3	1.1630; −6
horsepower hour	3.7251; −7	3.7251; −14	3.6530; −6	3.6530; −11	5.0505; −7	1.5697; −8	1.5596; −3	1.5596; −6
British thermal unit	9.4782; −4	9.4782; −11	9.2949; −3	9.2949; −8	1.2851; −3	3.9941; −5	3.9683; 0	3.9683; −3
electron volt	6.2414; +18	6.2414; +11	6.1208; +19	6.1208; +14	8.4623; +18	2.6302; +17	2.6132; +22	2.6132; +19

multiply by the factor in the appropriate cell of the table

kW h	hp h	Btu	eV	← To convert Into ↓
3.6000; +6	2.6845; +6	1.0551; +3	1.6022; −19	joule
3.6000; +13	2.6845; +13	1.0551, +10	1.6022; 12	erg
3.6710; +5	2.7374; +5	1.0759; +2	1.6338; −20	kilogramme-force metre
3.6710; +10	2.7374; +10	1.0759; +7	1.6338; −15	gramme-force centimetre
2.0662; +6	1.0800; +6	7.7017; +2	1.1017; 10	foot pound force
8.5429; +7	6.3705; +7	2.5037; +4	3.8021; −18	foot poundal
8.5985; +2	6.4119; +2	2.5200; −1	3.8268; −23	kilocalorie
8.5985; +5	6.4119; +5	2.5200; +2	3.8268; −20	calorie
1	7.4570; −1	2.9307; −4	4.4505; −26	kilowatt hour
1.3410; 0	1	3.9301; −4	5.9683; −26	horsepower hour
3.4121; +3	2.5444; +3	1	1.5186; −22	British thermal unit
2.2469; +25	1.6755; +25	6.5851; +21	1	electron volt

POWER ENERGY CONSUMPTION

Dimensions $M\,L^2\,T^{-3}$ or $F\,L\,T^{-1}$

SI Unit watt (W)

$1\,W = 1\,J\,s^{-1}$

To convert → Into ↓	W	kW	erg s⁻¹	kgf m s⁻¹	gf cm s⁻¹	CV	ft lbf s⁻¹	ft pdl s⁻¹
watt	1	1.0000; +3	1.0000; −7	9.8067; 0	9.8067; −5	7.3550; +2	1.3558; 0	4.2140; −2
kilowatt	1.0000; −3	1	1.0000; −10	9.8067; −3	9.8067; −8	7.3550; −1	1.3558; −3	4.2140; −5
erg per second	1.0000; +7	1.0000; +10	1	9.8067; +7	9.8067; +2	7.3550; +9	1.3558; +7	4.2140; +5
kilogramme-force metre per second	1.0197; −1	1.0197; +2	1.0197; −8	1	1.0000; −5	7.5000; +1	1.3825; −1	4.2971; −3
gramme-force centimetre per second	1.0197; +4	1.0197; +7	1.0197; −3	1.0000; +5	1	7.5000; +6	1.3825; +4	4.2971; +2
metric horsepower (cheval-vapeur)	1.3596; −3	1.3596; 0	1.3596; −10	1.3333; −2	1.3333; −7	1	1.8434; −3	5.7295; −5
foot pound-force per second	7.3756; −1	7.3756; +2	7.3756; −8	7.2330; 0	7.2330; −5	5.4248; +2	1	3.1081; −2
foot poundal per second	2.3730; +1	2.3730; +4	2.3730; −6	2.3272; +2	2.3272; −3	1.7454; +4	3.2174; +1	1
horsepower	1.3410; −3	1.3410; 0	1.3410; −10	1.3151; −2	1.3151; −7	9.8632; −1	1.8182; −3	5.6511; −5
kilocalorie per second	2.3885; −4	2.3885; −1	2.3885; −11	2.3423; −3	2.3423; −8	1.7567; −1	3.2383; −4	1.0065; −5
kilocalorie per minute	1.4331; −2	1.4331; +1	1.4331; −9	1.4054; −1	1.4054; −6	1.0540; +1	1.9430; −2	6.0390; −4
kilocalorie per hour	8.5985; −1	8.5985; +2	8.5985; −8	8.4322; 0	8.4322; −5	6.3242; +2	1.1658; 0	3.6234; −2
kilocalorie per day	2.0636; +1	2.0636; +4	2.0636; −6	2.0237; +2	2.0237; −3	1.5178; +4	2.7979; +1	8.6962; −1
British thermal unit per minute	5.6869; −2	5.6869; +1	5.6869; −9	5.5769; −1	5.5769; −6	4.1827; +1	7.7104; −2	2.3965; −3

multiply by the factor in the appropriate cell of the table

hp	kcal s^{-1}	kcal min^{-1}	kcal h^{-1}	kcal d^{-1}	Btu min^{-1}	← To convert Into ↓
7.4570; +2	4.1868; +3	6.9780; +1	1.1630; 0	4.8458; −2	1.7584; +1	watt
7.4570; −1	4.1868; 0	6.9780; −2	1.1630; −3	4.8458; −5	1.7584; −2	kilowatt
7.4570; +9	4.1868; +10	6.9780; +8	1.1630; +7	4.8458; +5	1.7584; +8	erg per second
7.6040; +1	4.2693; +2	7.1156; 0	1.1859; −1	4.9414; −3	1.7931; 0	kilogramme-force metre per second
7.6040; +6	4.2693; +7	7.1156; +5	1.1859; +4	4.9414; +2	1.7931; +5	gramme-force centimetre per second
1.0139; 0	5.6925; 0	9.4874; −2	1.5812; −3	6.5885; −5	2.3908; −2	metric horsepower (cheval-vapeur)
5.5000; +2	3.0000; +10	6.1107; +1	8.0770; +1	1.0141; −7	1.0000; +1	foot pound-force per second
1.7696; +4	9.9354; +4	1.6559; +3	2.7598; +1	1.1499; 0	4.1728; +2	foot poundal per second
1	5.6146; 0	9.3577; −2	1.5596; −3	6.4984; −5	2.3581; −2	horsepower
1.7811; −1	1	1.6667; −2	2.7778; −4	1.1574; −5	4.1999; −3	kilocalorie per second
1.0686; +1	6.0000; +1	1	1.6667; −2	6.9444; −4	2.5200; −1	kilocalorie per minute
6.4119; +2	3.6000; +3	6.0000; +1	1	4.1667; −2	1.5120; +1	kilocalorie per hour
1.5388; +4	8.6400; +4	1.4400; +3	2.4000; +1	1	3.6287; +2	kilocalorie per day
4.2407; +1	2.3810; +2	3.9683; 0	6.6139; −2	2.7558; −3	1	British thermal unit per minute

METABOLIC RATE   SPECIFIC POWER OUTPUT

Dimensions $L^2 T^{-3}$

SI Unit watt per kilogramme (W kg^{-1})
(1 W kg^{-1} = 1 m^2 s^{-3})

To convert →	W kg^{-1}	erg s^{-1} g^{-1}	kcal s^{-1} g^{-1}	kcal min^{-1} g^{-1}	kcal h^{-1} g^{-1}	kcal d^{-1} g^{-1}	kcal s^{-1} kg^{-1} (= cal s^{-1} g^{-1})
Into ↓							
watt per kilogramme	1	1.0000; −4	4.1868; +6	6.9780; +4	1.1630; +3	4.8458; +1	4.1868; +3
erg per second gramme	1.0000; +4	1	4.1868; +10	6.9780; +8	1.1630; +7	4.8458; +5	4.1868; +7
kilocalorie per second gramme	2.3885; −7	2.3885; −11	1	1.6667; −2	2.7778; −4	1.1574; −5	1.0000; −3
kilocalorie per minute gramme	1.4331; −5	1.4331; −9	6.0000; +1	1	1.6667; −2	6.9444; −4	6.0000; −2
kilocalorie per hour gramme	8.5985; −4	8.5985; −8	3.6000; +3	6.0000; +1	1	4.1667; −2	3.6000; 0
kilocalorie per day gramme	2.0636; −2	2.0636; −6	8.6400; +4	1.4400; +3	2.4000; +1	1	8.6400; +1
kilocalorie per second kilogramme	2.3885; −4	2.3885; −8	1.0000; +3	1.6667; +1	2.7778; −1	1.1574; −2	1
kilocalorie per minute kilogramme	1.4331; −2	1.4331; −6	6.0000; +4	1.0000; +3	1.6667; +1	6.9444; −1	6.0000; +1
kilocalorie per hour kilogramme	8.5985; −1	8.5985; −5	3.6000; +6	6.0000; +4	1.0000; +3	4.1667; +1	3.6000; +3
kilocalorie per day kilogramme	2.0636; +1	2.0636; −3	8.6400; +7	1.4400; +6	2.4000; +4	1.0000; +3	8.6400; +4
horsepower per pound	6.0828; −4	6.0828; −8	2.5467; +3	4.2446; +1	7.0743; −1	2.9476; −2	2.5467; 0

multiply by the factor in the appropriate cell of the table

$\text{kcal min}^{-1}\,\text{kg}^{-1}$ $(=\text{cal min}^{-1}\,\text{g}^{-1})$	$\text{kcal h}^{-1}\,\text{kg}^{-1}$ $(=\text{cal h}^{-1}\,\text{g}^{-1})$	$\text{kcal d}^{-1}\,\text{kg}^{-1}$ $(=\text{cal d}^{-1}\,\text{g}^{-1})$	hp lb^{-1}	← To convert
				Into ↓
6.9780; +1	1.1630; 0	4.8458; −2	1.6440; +3	watt per kilogramme
6.9780; +5	1.1630; +4	4.8458; +2	1.6440; +7	erg per second gramme
1.6667; −5	2.7778; −7	1.1574; −8	3.9266; −4	kilocalorie per second gramme
1.0000; −3	1.6667; −5	6.9444; 7	2.3560; −2	kilocalorie per minute gramme
6.0000; −2	1.0000; −3	4.1667; −5	1.4136; 0	kilocalorie per hour gramme
1.4400; 0	2.4000; −2	1.0000; −3	3.3926; +1	kilocalorie per day gramme
1.6667; −2	2.7778; −4	1.1574; −5	3.9266; −1	kilocalorie per second kilogramme
1	1.6667; −2	6.9444; −4	2.3560; +1	kilocalorie per minute kilogramme
6.0000; +1	1	4.1667; −2	1.4136; +3	kilocalorie per hour kilogramme
1.4400; +3	2.4000; +1	1	3.3926; +4	kilocalorie per day kilogramme
4.2446; −2	7.0743; −4	2.9476; −5	1	horsepower per pound

ENERGY FLUX   PRODUCTIVITY

Dimensions $M\,T^{-3}$ or $F\,L^{-1}\,T^{-1}$

SI Unit   watt per square metre $(W\,m^{-2})$
$(1\,W\,m^{-2} = 1\,kg\,s^{-3})$

Into ↓ To convert →	$W\,m^{-2}$	$cal\,cm^{-2}\,s^{-1}$	$cal\,cm^{-2}\,min^{-1}$	$kcal\,cm^{-2}\,s^{-1}$	$kcal\,cm^{-2}\,min^{-1}$	$kcal\,cm^{-2}\,h^{-1}$	$kcal\,m^{-2}\,s^{-1}$
watt per square metre	1	4.1868; +4	6.9780; +2	4.1868; +7	6.9780; +5	1.1630; +4	4.1868; +3
calorie per square centimetre second	2.3885; −5	1	1.6667; −2	1.0000; +3	1.6667; +1	2.7778; −1	1.0000; −1
calorie per square centimetre minute	1.4331; −3	6.0000; +1	1	6.0000; +4	1.0000; +3	1.6667; +1	6.0000; 0
kilocalorie per square centimetre second	2.3885; −8	1.0000; −3	1.6667; −5	1	1.6667; −2	2.7778; −4	1.0000; −4
kilocalorie per square centimetre minute	1.4331; −6	6.0000; −2	1.0000; −3	6.0000; +1	1	1.6667; −2	6.0000; −3
kilocalorie per square centimetre hour	8.5985; −5	3.6000; 0	6.0000; −2	3.6000; +3	6.0000; +1	1	3.6000; −1
kilocalorie per square metre second	2.3885; −4	1.0000; +1	1.6667; −1	1.0000; +4	1.6667; +2	2.7778; 0	1
kilocalorie per square metre minute	1.4331; −2	6.0000; +2	1.0000; +1	6.0000; +5	1.0000; +4	1.6667; +2	6.0000; +1
kilocalorie per square metre hour	8.5985; −1	3.6000; +4	6.0000; +2	3.6000; +7	6.0000; +5	1.0000; +4	3.6000; +3
kilocalorie per square metre day	2.0636; +1	8.6400; +5	1.4400; +4	8.6400; +8	1.4400; +7	2.4000; +5	8.6400; +4
kilocalorie per hectare day	2.0636; +5	8.6400; +9	1.4400; +8	8.6400; +12	1.4400; +11	2.4000; +9	8.6400; +8
British thermal unit per square foot second	8.8055; −5	3.6867; 0	6.1445; −2	3.6867; +3	6.1445; +1	1.0241; 0	3.6867; −1

multiply by the factor in the appropriate cell of the table

To convert →	$W\ m^{-2}$	$kW\ ha^{-1}$	$kW\ km^{-2}$	$W\ cm^{-2}$	$W\ mm^{-2}$	$mW\ mm^{-2}$	$erg\ s^{-1}\ cm^{-2}$	$erg\ s^{-1}\ mm^{-2}$
Into ↓								
watt per square metre	0	−1	−3	+4	+6	+3	−3	−1
kilowatt per hectare	+1	0	−2	+5	+7	+4	−2	0
kilowatt per square kilometre	+3	+2	0	+7	+9	+6	0	+2
watt per square centimetre	−4	−5	−7	0	+2	−1	−7	−5
watt per square millimetre	−6	−7	−9	−2	0	−3	−9	−7
milliwatt per square millimetre	−3	−4	−6	+1	+3	0	−6	−4
erg per second centimetre-squared	+3	+2	0	+7	+9	+6	0	+2
erg per second millimetre-squared	+1	0	−2	+5	+7	+4	−2	0

multiply by 10^n, where n is the number in the appropriate cell of the table

$kcal\ m^{-2}\ min^{-1}$	$kcal\ m^{-2}\ h^{-1}$	$kcal\ m^{-2}\ d^{-1}$	$kcal\ ha^{-1}\ d^{-1}$	$Btu\ ft^{-2}\ s^{-1}$	← To convert
					Into ↓
6.9780; +1	1.1630; 0	4.8458; −2	4.8458; −6	1.1357; +4	watt per square metre
1.6667; −3	2.7778; −5	1.1574; −6	1.1574; −10	2.7125; −1	calorie per square centimetre second
1.0000; −1	1.6667; −3	6.9444; −5	6.9444; −9	1.6275; +1	calorie per square centimetre minute
1.6667; −6	2.7778; −8	1.1574; −9	1.1574; −13	2.7125; −4	kilocalorie per square centimetre second
1.0000; −4	1.6667; −6	6.9444; −8	6.9444; −12	1.6275; −2	kilocalorie per square centimetre minute
6.0000; −3	1.0000; −4	4.1667; −6	4.1667; −10	9.7649; −1	kilocalorie per square centimetre hour
1.6667; −2	2.7778; −4	1.1574; −5	1.1574; −9	2.7125; 0	kilocalorie per square metre second
1	1.6667; −2	6.9444; −4	6.9444; −8	1.6275; +2	kilocalorie per square metre minute
6.0000; +1	1	4.1667; −2	4.1667; −6	9.7649; +3	kilocalorie per square metre hour
1.4400; +3	2.4000; +1	1	1.0000; −4	2.3436; +5	kilocalorie per square metre day
1.4400; +7	2.4000; +5	1.0000; +4	1	2.3436; +9	kilocalorie per hectare day
6.1445; −3	1.0241; −4	4.2670; −6	4.2670; −10	1	British thermal unit per square foot second

ENERGY CONTENT   SPECIFIC ENERGY

Dimensions $L^2 T^{-2}$

SI Unit   joule per kilogramme (J kg⁻¹)
$$(1\ J\ kg^{-1} = 1\ m^2\ s^{-2})$$

Notes   The *calorie* is the international table calorie.

Into ↓ / To convert →	J kg⁻¹	kgf m kg⁻¹	kgf m g⁻¹	gf cm g⁻¹	kcal kg⁻¹	kcal g⁻¹	ft lbf lb⁻¹	ft pdl lb⁻¹	Btu lb⁻¹
joule per kilogramme	1	9.8067; 0	9.8067; +3	9.8067; −2	4.1868; +3	4.1868; +6	2.9891; 0	9.2903; −2	2.3260; +3
kilogramme-force metre per kilogramme	1.0197; −1	1	1.0000; +3	1.0000; −2	4.2693; +2	4.2693; +5	3.0480; −1	9.4735; −3	2.3719; +2
kilogramme-force metre per gramme	1.0197; −4	1.0000; −3	1	1.0000; −5	4.2693; −1	4.2693; +2	3.0480; −4	9.4735; −6	2.3719; −1
gramme-force centimetre per gramme	1.0197; +1	1.0000; +2	1.0000; +5	1	4.2693; +4	4.2693; +7	3.0480; +1	9.4735; −1	2.3719; +4
kilocalorie per kilogramme (= calorie per gramme)	2.3885; −4	2.3423; −3	2.3423; 0	2.3423; −5	1	1.0000; +3	7.1393; −4	2.2190; −5	5.5556; −1
kilocalorie per gramme	2.3885; −7	2.3423; −6	2.3423; −3	2.3423; −8	1.0000; −3	1	7.1393; −7	2.2190; −8	5.5556; −4
foot pound-force per pound	3.3455; −1	3.2808; 0	3.2808; +3	3.2808; −2	1.4007; +3	1.4007; +6	1	3.1081; −2	7.7817; +2
foot poundal per pound	1.0764; +1	1.0556; +2	1.0556; +5	1.0556; 0	4.5066; +4	4.5066; +7	3.2174; +1	1	2.5037; +4
British thermal unit per pound	4.2992; −4	4.2161; −3	4.2161; 0	4.2161; −5	1.8000; 0	1.8000; +3	1.2851; −3	3.9941; −5	1

multiply by the factor in the appropriate cell of the table

DYNAMIC VISCOSITY

Dimensions $M L^{-1} T^{-1}$ or $F T L^{-2}$

SI Unit pascal second (Pa s)
$1\ Pa\ s = 1\ kg\ m^{-1}\ s^{-1} = 1\ N\ s\ m^{-2}$

Notes The form of the dimensions is such that several commonly-used units of dynamic viscosity can be expressed in two alternative forms. The following are alternative names for the same units:

$1\ poise = 1\ g\ cm^{-1}\ s^{-1} = 1\ dyn\ s\ cm^{-2}$

$1\ lb\ ft^{-1}\ s^{-1} = 1\ pdl\ s\ ft^{-2}$

$1\ sl\ ft^{-1}\ s^{-1} = 1\ lbf\ s\ ft^{-2}$

To convert → Into ↓	Pa s	P	kgf s m^{-2}	gf s cm^{-2}	lb ft^{-1} s^{-1}	lb in^{-1} s^{-1}	sl ft^{-1} s^{-1}	lbf s in^{-2}
pascal second	1	1.0000; −1	9.8067; 0	9.8067; +1	1.4882; 0	1.7858; +1	4.7880; +1	6.8948; +3
poise	1.0000; +1	1	9.8067; +1	9.8067; +2	1.4882; +1	1.7858; +2	4.7880; +2	6.8948; +4
kilogramme-force second per metre-squared	1.0197; −1	1.0197; −2	1	1.0000; +1	1.5175; −1	1.8210; 0	4.8824; 0	7.0307; +2
gramme-force second per centimetre-squared	1.0197; −2	1.0197; −3	1.0000; −1	1	1.5175; −2	1.8210; −1	4.8824; −1	7.0307; +1
pound per foot second	6.7197; −1	6.7197; −2	6.5898; 0	6.5898; +1	1	1.2000; +1	3.2174; +1	4.6331; +3
pound per inch second	5.5997; −2	5.5997; −3	5.4915; −1	5.4915; 0	8.3333; −2	1	2.6812; 0	3.8609; +2
slug per foot second	2.0885; −2	2.0885; −3	2.0482; −1	2.0482; 0	3.1081; −2	3.7297; −1	1	1.4400; +2
pound-force second per inch-squared	1.4504; −4	1.4504; −5	1.4223; −3	1.4223; −2	2.1584; −4	2.5901; −3	6.9444; −3	1

multiply by the factor in the appropriate cell of the table

KINEMATIC VISCOSITY

Dimensions $L^2 T^{-1}$

SI Unit square metre per second ($m^2\ s^{-1}$)

Notes All the units commonly used for expressing kinematic viscosity are based on the second as the unit of time, and differ only in the units of area used. Conversion factors can therefore be obtained from the table of area units (p 10). For instance, the factor for converting $ft^2\ s^{-1}$ into $m^2\ s^{-1}$ is the same as that for converting ft^2 into m^2.

$1\ stokes = 1\ cm^2\ s^{-1}$

PLANE ANGLE

Dimensions   None
SI Unit   radian (rad)

To convert → Into ↓	rad	rev	°	′	″	g
radian	1	$2\pi = 6.2832$; 0	$\dfrac{\pi}{180} = 1.7453$; -2	$\dfrac{\pi}{10800} = 2.9089$; -4	$\dfrac{\pi}{648000} = 4.8481$; -6	$\dfrac{\pi}{200} = 1.5708$; -2
revolution (= cycle)	$\dfrac{1}{2\pi} = 1.5915$; -1	1	2.7778; -3	4.6296; -5	7.7160; -7	2.5000; -3
degree	$\dfrac{180}{\pi} = 5.7296$; $+1$	3.6000; $+2$	1	1.6667; -2	2.7778; -4	9.0000; -1
arc minute	$\dfrac{10800}{\pi} = 3.4377$; $+3$	2.1600; $+4$	6.0000; $+1$	1	1.6667; -2	5.4000; $+1$
arc second	$\dfrac{648000}{\pi} = 2.0626$; $+5$	1.2960; $+6$	3.6000; $+3$	6.0000; $+1$	1	3.2400; $+3$
grade	$\dfrac{200}{\pi} = 6.3662$; $+1$	4.0000; $+2$	1.1111; 0	1.8519; -2	3.0864; -4	1

multiply by the factor in the appropriate cell of the table

TEMPERATURE

To convert →	deg K	deg C	deg R	deg F
Into ↓				
degree Kelvin	$y = x$	$y = x + 273.15$	$y = \dfrac{5}{9} x$	$y = \dfrac{5}{9} (x + 459.67)$
degree Celsius	$y = x - 273.15$	$y = x$	$y = \dfrac{5}{9} (x - 491.67)$	$y = \dfrac{5}{9} (x - 32)$
degree Rankine	$y = \dfrac{9}{5} x$	$y = \dfrac{9}{5} x + 491.67$	$y = x$	$y = x + 459.67$
degree Fahrenheit	$y = \dfrac{9}{5} x - 459.67$	$y = \dfrac{9}{5} x + 32$	$y = x - 459.67$	$y = x$

set x equal to the temperature in the given units, to obtain y, the same temperature in the required units.

INDEX

This index contains a complete list of the units in the tables, but in the case of composite units, only one out of a group of related units is listed in full. This is followed by a list of the other units, indicated by the element(s) in which they differ from the main one, thus — kilocalorie per second (*minute, hour, day*). The index is intended to serve also as a glossary, from which the nature of an unfamiliar unit can be ascertained.

SI Units are listed in heavy type.

Unit name	Abbreviation	Type of quantity	Page	Unit name	Abbreviation	Type of quantity	Page
dyne per square centimetre (*sq millimetre*)	dyn cm^{-2}	pressure	26	gramme per square centi-metre (*sq metre, sq millimetre*)	g cm^{-2}	biomass density	20
dyne second per centimetre-squared (= poise)	P	dynamic viscosity	37	gramme per square foot	g ft^{-2}	biomass density	20
				gramme per cubic centi-metre (*cu metre*)	g cm^{-3}	density	18
electron volt	eV	energy	28				
erg	erg	work	28	gramme per centimetre-second (= poise)	P	dynamic viscosity	37
erg per second	erg s^{-1}	power	30				
erg per second gramme	erg s^{-1} g^{-1}	metabolic rate	32	gramme-force	gf	force	23
erg per second centimetre-squared (*millimetre-squared*)	erg s^{-1} cm^{-2}	energy flux	34	gramme-force per centi-metre (*millimetre*)	gf cm^{-1}	membrane tension	24
				gramme-force per square centimetre (*sq millimetre*)	gf cm^{-2}	pressure	26
fathom	fath	length	8				
fluid ounce, UK	UK fl oz	volume	12	gramme-force centimetre	gf cm	work	28
fluid ounce, US	US fl oz	volume	12	gramme-force centimetre per gramme	gf cm g^{-1}	energy content	36
foot	ft	length	8				
foot of water	ft H$_2$O	pressure	26	gramme-force centimetre per second	gf cm s^{-1}	power	30
foot per second (*minute*)	ft s^{-1}	speed	15				
foot poundal	ft pdl	work	28	gramme force second per centimetre-squared	gf s cm^{-2}	dynamic viscosity	37
foot poundal per second	ft pdl s^{-1}	power	30				
foot poundal per pound	ft pdl lb^{-1}	energy content	36	hectare	ha	area	10
foot pound-force	ft lbf	work	28	horsepower	hp	power	30
foot pound-force per second	ft lbf s^{-1}	power	30	horsepower hour	hp h	work	20
				horsepower per pound	hp lb^{-1}	metabolic rate	32
foot pound-force per pound	ft lbf lb^{-1}	energy content	36	horsepower, metric	CV	power	30
furlong	fg	length	8	hour	h	time	14
gallon, UK	UK gal	volume	12	hundredweight	cwt	mass	16
gallon, US	US gal	volume	12	inch	in	length	8
grade	g	angle	38	inch of mercury	in Hg	pressure	26
gramme	g	mass	16	inch of water	in H$_2$O	pressure	26
gramme centimetre-squared	g cm^2	moment of inertia	22	international nautical mile	NM	length	8

Unit name	Abbreviation	Type of quantity	Page	Unit name	Abbreviation	Type of quantity	Page
pound-force second per inch-squared (*foot-squared*)	lbf s in^{-2}	dynamic viscosity	37	square yard	yd^2	area	10
				standard atmosphere	atm	pressure	26
				statute mile	mile	length	8
poundal	pdl	force	23	stokes	St	kinematic viscosity	37
poundal per foot	pdl ft^{-1}	membrane tension	24	stone	st	mass	16
poundal per square foot	pdl ft^{-2}	pressure	26	ton, short	sh ton	mass	16
poundal second per foot-squared	pdl s ft^{-2}	dynamic viscosity	37	ton, UK (= long ton)	UK ton	mass	16
				ton-force	tonf	force	23
radian	**rad**	**angle**	38	ton-force per foot (*inch, yard*)	tonf ft^{-1}	membrane tension	24
revolution	rev	angle	38				
rod (= pole)	rod	length	8	ton-force per square foot (*sq inch, sq yard*)	tonf ft^{-2}	pressure	26
rood	rood	area	10				
second	**s**	**time**	14	tonne	t	mass	16
short ton	sh ton	mass	16	tonne per hectare (*sq kilometre*)	t ha^{-1}	biomass density	20
sidereal day	si d	time	14				
sidereal month	si mon	time	14	tonne-force	tf	force	23
slug	sl	mass	16	tonne-force per metre (*centimetre*)	tf m^{-1}	membrane tension	24
slug per cubic foot (*cu inch, cu yard*)	sl ft^{-3}	density	18				
				tonne-force per square metre (*sq centimetre*)	tf m^{-2}	pressure	26
slug foot-squared	sl ft^2	moment of inertia	22				
slug per foot second	sl ft^{-1} s^{-1}	dynamic viscosity	37	UK fluid ounce	UK fl oz	volume	12
square centimetre	cm^2	area	10	UK pint	UK pt	volume	12
square chain	ch^2	area	10	UK gallon	UK gal	volume	12
square foot	ft^2	area	10	UK ton (= long ton)	UK ton	mass	16
square inch	in^2	area	10	UK ton per acre (*sq mile*)	UK ton acre^{-1}	biomass density	20
square kilometre	km^2	area	10	UK ton per cubic yard	UK ton yd^{-3}	density	18
square metre	**m^2**	**area**	10	US fluid ounce	US fl oz	volume	12
square mile	mile^2	area	10	US pint	US pt	volume	12
square millimetre	mm^2	area	10	US gallon	US gal	volume	12
square nautical mile	NM^2	area	10	**watt**	**W**	**power**	30
square rod	rod^2	area	10	**watt per kilogramme**	**W kg^{-1}**	**metabolic rate**	32

Unit name	Abbreviation	Type of quantity	Page
watt per square metre (*sq centimetre, sq millimetre*)	W m^{-2}	energy flux	34
week	wk	time	14
yard	yd	length	8
year	y	time	14

FURTHER CONVERSION FACTORS

To convert	Into	Multiply by	Notes

To convert	Into	Multiply by	Notes